HYDRO-PLOWING THE OCEAN

or

SOLVING GLOBAL WARMING AND WORLD HUNGER FOR FUN AND
PROFIT

By William Eugene Malley III G.G.F.

Copyright 2011. All rights reserved

Table of Contents

Part 1;

Chapter 1; Saving the Humans

Americans, as a general rule, aren't easily stampeded into anything. As far as the majority of Americans are concerned, human caused global warming doesn't exist. When the sleeping giant is once again awakened, the world will probably rejoice. Americans have always had a reputation of getting big jobs done. Many countries have already started halfhearted, ineffective, attempts to get global warming under control.

In this book we will show the world how to simply stop, and reverse, further global warming. The American method will also benefit most of the starving people in the world. A dual purpose crusade is always an American favorite. Americans are famous for not ever wanting anyone to go hungry.

There's one more thing that Americans enjoy doing. We like to thumb our collective noses, at the rest of the world, whenever the rest of the world decides to try and tell us what we should, or shouldn't, be doing.

We live on an incredibly beautiful blue planet. Unfortunately our planet is in danger. Our oceans are sick and quickly becoming terminal. It's been said that overfishing; pollution; and global warming are the culprits. The way the oceans go, so goes the earth.

For tens of thousands of years we were hunter gatherers on land, as well as the oceans. We eventually settled down and began the cultivation of the land. But when it comes to the oceans we're still nothing more than hunter gatherers. It's now time to begin the serious cultivation of the oceans. When we began growing crops on land we finally had a food supply we could depend on. The same is true of the oceans. As simple hunter gatherers, we are currently overfishing our oceans.

We are only engaging in overfishing because it is necessary to sustain the lives of millions of human beings. If there were a lot more fish we wouldn't be overfishing. This book will examine how global seafood supplies can actually be tripled or quadrupled, and at the same time leave the oceans, and our atmosphere, healthier than they've ever been.

This plan will reduce ocean pollution to a pre-industrialization level. At the same time these methods will turn billions of tons of carbon dioxide into oxygen. The solution to these problems will be much cheaper and easier then most people believe. Not only are the problems solvable, solving them will produce a great deal of financial profit for any of us who wish to participate.

This book lays out a plan, that when implemented, will bring our planet back to a healthy vigorous status that hasn't been seen in over two hundred years. This plan will save the lives of millions of starving people; most of them children. At the same time we will be turning hundreds of millions of tons of carbon dioxide into clean fresh oxygen. It's not a win-win situation, it's a win, win, win, win, situation.

It strikes me as a bit humorous that the dominant animal species on the planet Earth, will depend on the smallest plant in our lakes and oceans to save its future. The humble phytoplankton will be our tireless champion.

This book will discuss phytoplankton and zooplankton. We will discover how to safely increase their numbers at least tenfold. We will, at the same time, increase the amount of fish and seafood available by the same numbers. As the phytoplankton grows, it will remove hundreds of millions of tons of carbon dioxide from our atmosphere. We will finally be cultivating the other 129 million square miles of our planet.

We won't introduce any chemicals or fertilizer into the oceans to achieve this miracle. We will only use what's already there. In essence, we will be moving nature's own nutrients, from where they are, to where they are needed.

If we were selling this phytoplankton and zooplankton by

the pound, we could label it 100% organic.

These microscopic plants and animals consist of many types. There are as many species of phytoplankton (Microscopic plants) in the ocean, and our lakes, as there are different kinds of plants on land.

As you probably know, all plants sequester large amounts carbon dioxide. What most people don't know, is that the oceans contain 85% of the plant life on planet Earth. Our lakes and rivers probably account for another 5%.

Think about that for a moment. All of the forests; jungles; shrubs; and grasses on land, only make up 10% of our planetary plant life.

There are as many species of zooplankton (Tiny animals) in the ocean, and our lakes, as there are different kinds of insects on land.

These different species of tiny plants and animals reproduce in many different ways. Their methods of reproduction aren't of much importance; just let it be known, that under the proper conditions they are very very prolific.

Phytoplankton are the foundation of the aquatic food chain. To reduce the excess carbon dioxide in our atmosphere and increase our seafood supply we only need to stimulate the natural growth of phytoplankton.

One square-mile of properly managed phytoplankton can permanently sequester as much carbon dioxide as 50 square miles of forest. You must understand that eventually the trees of a forest will, in one way or another, be burned. The burning could take place because of a forest fire, or when products built from the wood of these forests reach the end of their useful life. The burning of these forest products will reintroduce the forest sequestered carbon dioxide into the atmosphere.

This is definitely not a problem with phytoplankton. Once the majority of phytoplankton die, they drop safely to the bottom of the sea. The remains of the phytoplankton will lay on the bottom of the ocean virtually forever.

A moderate percentage of the phytoplankton will be consumed by what we humans usually refer to as fish, or seafood. By accelerating the growth of phytoplankton we are helping in the production of millions of tons of tasty protein. Along the way we will get rid of hundreds of millions of tons of carbon dioxide.

You may not think that you consume very much fish, or seafood. But in today's world that's usually not true. If you eat organic fruits and vegetables you might be surprised to learn that a major fertilizer used on organic fruits and vegetables is fish fertilizer. This fish fertilizer is usually made from what is left over after seafood is processed.

If you are a true carnivore you might also be surprised to learn that some of the feed that most domestic animals consume, contains fish-meal, as a protein supplement. Almost everything we eat has ties back to the lowly phytoplankton. Life would become much harder for everyone without this small insignificant plant. As you continue to read this book you will discover that the health of these tiny plants is vitally import to almost everyone on the planet.

If you mix these three simple things together correctly, you get incredible phytoplankton growth. The three things you mix together are, sunlight, carbon dioxide, and nutrient rich water.

In winter we have enough of two of these three things. We have a lot of nutrients, plenty of carbon dioxide, but not really as much sunlight as we could use.

In summer we have plenty of sunlight, and carbon dioxide, but the phytoplankton quickly use up most of the nutrients in the sunlight zone. (sunlight zone = top 40 to 70 ft. of bodies of water) Most phytoplankton will only grow in the sunlight zone. So in summer, because of nutrient depletion, phytoplankton production drops to about one 50th of what it should be.

In this portion of the book we will concentrate on our oceans. Two later chapters will focus on lakes.

Phytoplankton are very important to us, this tiny plant has a dramatic affect on the worldwide seafood supply, as well as carbon dioxide levels.

How does phytoplankton affect that can, or packet, of tuna in your cupboard? Can it actually affect the nice filet of sole in your refrigerator?

YES!

That can, or packet, of tuna on your shelf, and the filet of sole in your refrigerator, wouldn't exist if it wasn't for phytoplankton. There would be virtually no life in the ocean if it weren't for those same small plants.

Large fish eat plankton?

In most cases not directly, here's how it usually works.

The phytoplankton, a very primitive plant, lays around in the top 40 or 50 ft. of the ocean, sometimes reproducing like crazy. Along come tiny creatures called Zooplankton. These are the smallest ocean creatures that can be loosely classified as animals. These tiny animals live by devouring the phytoplankton. Zooplankton are often consumed by slightly larger creatures. Some of these tiny animals are eaten by larger animals, until we get up to the small creatures that the large fish eat.

Some of the largest creatures on earth consume mostly zooplankton. The zooplankton known as krill, is the main food of several species of whales. Almost all shellfish and coral live on either phytoplankton or zooplankton.

With a few exceptions, it's the same story with all of the other food chains, it all starts with the phytoplankton.

Chapter 2; Understanding the Natural Process

I have always held that the carbon problems (carbon dioxide) that we are having are part of a cycle. If we wish to remove vast amounts of carbon dioxide (or carbon) from the atmosphere, the simplest and most cost-effective way to do this would be to tap into, and accelerate, the natural processes that cycle carbon through our biosphere. By doing this we will derive many more benefits than just the removal of the carbon dioxide. We will provide vast amounts of food for people all over the world.

Carbon dioxide is a molecule made up of one atom of carbon and two atoms of oxygen. If we remove the one atom of carbon we are left with two atoms of oxygen (02). So the problem isn't the removal of carbon dioxide. The real problem is; the removal of that one carbon atom in each carbon dioxide molecule. This this is easily achieved using phytoplankton photosynthesis.

Each day millions of tons of Carbon dioxide are removed from the atmosphere by the natural growth of algae or plankton in our lakes and oceans. All we need do to remove many more millions of tons of carbon dioxide from the atmosphere is to incite unusual phytoplankton growth in relatively small areas. These tiny plants, once they complete their life-cycle, fall to the ocean, or lake bottom, carrying their carbon with them. At this point, the plankton is approximately 40% carbon (dry weight). The carbon stays on the bottom for thousands or even millions of years.

This carbon is in a very stable state. It's not going to suddenly turn back into carbon dioxide and leak back into the atmosphere, unlike some of the silly schemes for the removal and storage of carbon dioxide. There's absolutely no chance of it ever becoming a problem again. The carbon dioxide molecule has been permanently disassembled. The only way this ocean stored carbon could turn back into carbon dioxide, is if we pulled it up from the bottom and burned it. The oil that we burn so recklessly started off as algae or plankton hundreds of thousands, or millions of years ago. It went through its life-cycle, settled to the bottom of the ocean, was covered with sentiment, and over time became oil.

So in essence, what we're doing is using solar energy (Photosynthesis) to convert the carbon dioxide produced by the oil, and coal, that we burn, back into a stable form of carbon. Then nature stores it back on the seafloor. This is a truly elegant solution to the situation we find ourselves in.

The world's oceans regularly absorb and expel atmospheric gases. You might say the oceans breathe. This absorption and releasing of atmospheric gases is dependent on many factors, wind speed, temperature, etc. The oceans breathe much more oxygen out, then they breathe in.

It could be correctly stated that our seas are the world's main oxygen factories. With over 120 million square miles of ocean this fact isn't in dispute. The oceans manage this magic trick by breaking carbon dioxide gas down into carbon and oxygen. The oceans themselves don't actually do this, the small plants that grow in the ocean are the actual workers. This isn't really a magic trick, it's just the photosynthesis engaged in by all plants.

Phytoplankton are the main gears in the global oxygen machine. As stated before, the plankton absorbs carbon dioxide gas, it then breaks the gas down, releasing the oxygen and retaining the carbon within itself.

In most cases the majority of plankton growth in the oceans occurs in the winter. It would seem to the average person, that in summer the growing conditions for plankton would be much better, more light, warmer. The average person would be absolutely correct, except for one little fly in the ointment. Summer ocean thermal stratification.

In winter, the surface (The top 40 or 50 feet) of the ocean is constantly re-mineralized by natural convection. In the winter the surface of the ocean usually cools to about the same temperature as the depths. At this point something called the thermocline breaks up and disappears. (More about the thermocline in a few moments.) This allows the deep water to upwell and mix with the surface water. This natural upwelling provides the algae with a continuous winter supply of nutrients. Unfortunately these nutrients are, to a great extent, wasted. The low light levels, and the cold, restrict the growth of the phytoplankton.

A few weeks into spring the oceans begin to stratify, a thermal boundary layer begins to appear. The thermal boundary layer is characterized by warm water above, and cold water below. This is usually referred to as the thermocline. This thermal boundary layer begins to form a few inches below the surface, and over several weeks moves down to 60 or 70 ft. The thermocline stops the mixing of the cold, more dense, nutrient rich, deep water, with the warmer, less dense, surface water as effectively as a sheet of incredibly thin plastic. This cuts off the normal updraft of nutrients from the depths. At this point, phytoplankton start growing like crazy.

Within three or four weeks the microscopic plants use up

the existing minerals in the sunlight zone, and then begin to starve. This little, or no growth situation continues until mid autumn.

Chapter 4; That Dastardly Thermocline

Our first reaction to the thermocline might be to condemn it. But this thermal boundary layer, in essence, protects us from another ice age. If the ocean thermocline were to disappear completely, within a few years the Earth would be shrouded in ice. Hundreds of billions of tons of carbon dioxide would disappear from our atmosphere in a very short time. This would have the opposite effect of global warming, global cooling. What we can do, is very carefully mitigate some of the effect of the thermal boundary layer. So this book's solution to the carbon problem is to bring nutrient rich seawater to the surface of the ocean, from below the thermocline. This would usually only be done in summer. On the next page I have included calculations supposing a moderate induced algae bloom over a relatively small area. In the example I have presuppose the use of mixing boats. (More on mixing boats later.)

Chapter 5; Creating a safe algae bloom

Mixing boats;

Mixing boats would be nothing more than 30 to 60 ft. ocean going boats, or ships. These vessels would be equipped with long flexible hoses, and pumps, that would pull nutrient rich water from below the thermocline, and spray it on the surface. This action would fertilize the sunlight zone with the minerals that the phytoplankton so desperately need.

Timeframe 20 to 25 days

1 sq mile = 27.8 million square feet. X 1 lb of algae per square foot = 27.8 million lbs of algae. (Dry weight)

40% carbon content of algae equals 11 million lbs of carbon per square mile. Or 5,500 tons.

140 miles X 100 miles artificial algae bloom = 14,000 sq miles.

5,500 tons of carbon per square mile X 14,000 sq miles = 77 million Tons of carbon removed from the atmosphere in 20 to 25 days.

The estimate of 1 pound of dry algae per square foot may seem to some to be an extravagant claim. But you must take into consideration that each square foot of ocean surface is actually the top of a very tall column of seawater. Phytoplankton plants only live a day or two before dying and falling to the bottom of the sea. We must remember that the microscopic plants propagate at a very high rate. Under proper conditions most phytoplankton double in number each day. This estimate only covers a 20 to 25 day period. I estimate that 20 to 25 days would be the length of time that the re-mineralization of the upper ocean would last. This assumes a single spot treatment of the given area. After about 25 days the plankton would have probably have consumed most of the newly introduced nutrients.

Chapter 6 ; Phytoplankton and Pollution

some interesting developments have occurred recently. It's been discovered that pollution actually helps phytoplankton grow. Phytoplankton react to fertilizer pollutants exactly the way a Forest, or lawn responds, it grows faster. It's been found that pollution helps in the absorption of carbon dioxide. As the phytoplankton grows faster, and reproduces sooner, it causes more carbon to be absorbed and carried to the ocean floor.

This means as it is carrying carbon to the sea floor, phytoplankton is also taking pollutants with it. It has been estimated that the carbon dioxide levels would be almost double what they are now, if phytoplankton had not been doing its job since the earliest days of industrialization. Each time we burn a lump of coal, or a gallon of gasoline, we're helping phytoplankton grow faster somewhere in the world.

There seems to be a fear that the oceans are becoming acidic. It's believed, as lakes and oceans become acidic they are causing calcium carbonate to percolate out of the water. Calcium carbonate is necessary for the production of the protective shells needed by many forms of aquatic life.

If you mix a lot of carbon dioxide with water you come up

with carbolic acid. People who should be in the know, believe that this acidity will eventually destroy the coral reefs. I disagree with this theory. Calcium carbonate is an incredibly plentiful material. The sheet-rock walls in your living room are mostly calcium carbonate. To think that there could ever be a shortage of calcium carbonate in the oceans is absolute foolishness.

Not too long ago I was watching a television program where they were examining what they said was a dying coral reef. They went on and on about how man-made pollution was destroying our coral reefs and would eventually destroy the ocean. As I was watching I noticed something, it was incredible how crystal clear the water was. There was absolutely no cloudiness to the water at all. The water seemed to be cleaner than the cleanest swimming pool. So I have a question for you. I will give you two hints. Hint Number One; most coral are filter feeders. (They live on plankton) Hint Number Two; As I said before, the water was crystal clear.

The question is; was man-made pollution really causing the damage to the coral reef? If not what was?

You probably guessed right, if you didn't I'm ashamed of you. The answer, is of course, there was no plankton in the water, the coral was starving to death. If you didn't get the correct answer, you need to go back to the beginning of this book and start over.

Once again, the best way to remove the carbon dioxide from our lakes and oceans is to increase the production of algae and phytoplankton. No matter how you you look at these problems, the question always seems to come back to the same answer. Grow more algae and phytoplankton. I think by this point you should be schooled enough to understand and agree with me.

Sand-eels are the main prey of many seabirds. Recently many of these seabirds have failed to reproduce. So what do sand-eels have to do with the reproduction of seabirds, and what does any of that have to do with plankton?. Sand-eels live on plankton, and the plankton seems to have moved. The plankton that the sand-eels live on have moved north; and the sand-eels have followed. This movement left the sea birds behind, with much less to eat.

Now we know that phytoplankton doesn't move; it's a free-floating plant. Once again the villain seems to be global warming. The feeding areas of the seabirds have warmed, and the thermocline is forming earlier and ending later. This is a case of phytoplankton underproduction affecting seabirds. Once again the answer is the same; "Break the Thermocline."

Chapter 7; The Incredible Benefits

50% of the worldwide legal fish harvest comes from approximately 1,294,000 square miles of the world's oceans. These small rich fisheries are scattered around the planet, and are known as natural areas of upwell.

These upwell areas sound like they encompass a large area until you understand that if they were all brought together, they would only be a section ocean of about 1000 miles by 1294 miles. About 1% of surface of all oceans. (Less then half the surface area of Canada) Not a very large piece when you calculate that all of the oceans comprise an area of over 129 million square miles.

Upwell areas are areas that naturally mix the nutrients from the depths of the ocean with the warmer, nutrient poor, surface waters, in summer, as well as in winter.

These upwell areas can be the result of worldwide deep-water ocean currents. As these currents flow around the planet, occasionally they'll move across an area of shallower water and are forced up to, or near, the surface. When this happens these currents provide the nutrients that the surface phytoplankton require. (In summer they naturally defeat the thermocline) I'm sure everyone has heard about the incredible commercial fishing on the Grand Banks.

The Grand Banks is one of these areas where a current is forced up to the surface, thus inducing extreme phytoplankton growth year-round. The fish are drawn to this abundance, and that's why so many commercial fishing boats ply these waters.

Another type of upwell is caused by seasonal coastal offshore winds, usually in the summer, off the coast of Africa. As the offshore wind blows the warm surface water out to sea, cooler water from deeper in the ocean rises up to fill this surface deficit. This natural summer upwelling provides an abundance of nutrients for phytoplankton, which are then preyed on by zooplankton, krill, etc. These microscopic plants and animals are eaten by larger sea creatures, who in turn are eaten by the fish that we harvest.

These natural areas of upwelling are vital to existing fish stocks. While these areas are providing food for ocean denizens, they're also carrying approximately 7,000,000,000 tons of carbon to the seafloor each month. The carbon removal for each year would total approximately 80,000,000,000 tons. This means that these areas clean the atmosphere of between 140,000,000,000 and 160,000,000,000 tons of carbon dioxide each year

If we were able to induce artificial upwell areas equal to the natural upwell areas, the Virgin Earth Challenge goal of 1,000,000,000 tons would be met in less than four days. There would be an incredible side benefit. This side benefit would make the endeavor worthwhile based on its merit alone. As well as consuming carbon dioxide and, in most cases, falling to the bottom of the sea, the plankton would provide incredible amounts of food for the ocean's ecosystem.

In five years, we would increase the available fish stocks on the planet by at least 50%. If properly managed, over time, we could increase the seafood output of the oceans by as much as 10 times. Tens of thousands of fishermen would go back to work, and millions of people would get the inexpensive protein they so desperately need.

The main reason for removing carbon dioxide from the atmosphere is to slow or stop global warming. The artificial increase of phytoplankton would accomplish this in a newly discovered manner. NASA funded scientists (Dierdre Toole of the Woods Hole Oceanographic Institution and David Siegel of the University of California, Santa Barbara) have discovered that phytoplankton, when exposed to high levels of sunlight (UV) produce a chemical compound called Dimethylsulfoniopropionate (DMSP).

This chemical is then broken down into Dimethylsulfide (DMS) by bacteria. DMS then moves from the ocean into the atmosphere, where it changes into several different sulfur compounds. Sulfur compounds in the DMS stick together in the atmosphere and create tiny dust particles. These particles of sulfurous dust are precisely the right size and shape for water to condense around, which is the usual manner in which clouds are formed. So, indirectly, plankton help create more clouds, and these clouds lower the Earth's surface temperature by reflecting some of the sun's energy back into space.

As many people approach the ocean on a sunny day, they notice that there comes a point where they believe they can actually smell the ocean. These sulfur compounds are what they are usually smelling.

Part 2;

Chapter 8; There Are Many Ways to Feed the Plankton

Basically all you need do to fertilize the nutrient poor upper areas of the ocean is to move nutrient rich deep seawater to the surface. This seawater must come from a depth of at least 100 feet.(The deeper the better) You are not actually using all that much energy to achieve this, because in essence, what you're doing is moving liquid around in a liquid environment.

Once it flows to the surface most of the energy or fuel involved would be used to disperse it. The best way to move the water from 100 feet below would be to use a tube, or flexible hose, that goes down to at least that depth. (Deep seawater well) As you remove water from the top of the tube, or hose, the deep water naturally flows up. Depending on the power source, you would then disperse this cooler sea water on the surface, in as large a pattern as practical. You couldn't just dump the nutrient rich deep water over the side, because the colder sea water, resists mixing with the warmer surface seawater, and most of it would fall back to the depths without providing much benefit.

Chapter 9; The Deep Seawater Well

Some people seem to have a problem understanding how the deep seawater well works; so I'm going to over-explain it a bit. My apologies to those of you who grasp it immediately.

In many parts of the the world they have a rather unique type of well. They are called Artesian wells. The beauty of an Artesian well is that the water in this type of well doesn't need to be pumped up to the surface. Once the well is drilled, the water flows to the surface on its own. This, to a certain degree, this is the same way the deep sea water well work's.

The nice thing about the deep seawater well, is that it doesn't have to be drilled. The deep seawater well consists of one end of a 100+ ft long, pipe, or tube, thrust into the ocean to a depth of at least 100 ft. (Below the thermocline)

This tube, or pipe, is open at both ends. The top of the tube, or pipe, is held high enough above the surface to prevent the entry of surface seawater. (3 ft. to 5 ft.) As you remove water from the top of this well, deep seawater naturally flows into the bottom, and up to the surface. After a few minutes of pumping, the deep seawater is at the surface of the well. No further pumping is necessary.

You have nutrient rich, deep seawater, available free for the taking, right there at the surface. This means that absolutely no further energy is needed to bring as much of the deep, nutrient rich, seawater to the surface as you might need, just like an artesian well.

The tube or pipe can be made of just about any material. The best choice would be a material that would retard, or resist clogging by marine life.

This tube could be rigid, or flexible enough to be rolled up. It could be as strong as quarter inch stainless steel, or as thin as the tubing you use to hook up your clothes dryer at home. The diameter of the tube, or pipe, could be anywhere from a few inches, to several feet. I'm sure the final choice would fall somewhere between these extremes.

If we create a deep seawater well with the diameter of 3 ft. we could actually use a bucket to lift as much of the nutrient rich deep seawater out of the surface of the well as we need. As we take water out of the top of the well, new, nutrient rich, deep seawater, naturally flows into the bottom, on its way up to the surface.

Picture for a moment a 3 ft. diameter 100 ft. long plastic

pipe. You are on a boat, 4 or 5 mi. out on the ocean. The pipe has been lowered vertically into the water to a point where you can only see the top 3 ft. of of it. There's water all around your boat, and of course there's water inside the pipe.

The surface of the water around you, and the level of the water inside the pipe is exactly the same. You put a hose into the water at the top of the pipe and begin pumping water out of it. The level of water in the pipe stays exactly the same. How can this be?

When you pump water out of the top of this pipe there's only one place new water can come in to replace it. 100 ft. below, nutrient rich deep seawater is flowing into the bottom, and soon it will reach you at the top.

I'm sure at this point the vast majority of you understand how a deep seawater well works. Once again my apologies to those of you who understood it quickly.

REVIEW I

At this point you should understand The Following.

1. There are as many species of phytoplankton in the ocean as there are different kinds of plants on land.

2. There are as many species of zooplankton in the ocean as there are different kinds of insects on land.

3. We are only engaging in overfishing because it is necessary to sustain the lives of millions of human beings. This book how global fishing can be tripled or quadrupled, and at the same time leave the oceans, and our atmosphere healthy.

4. Phytoplankton are necessary for the existence of all of the ocean's creatures.

5. We won't introduce any chemicals or fertilizer into the oceans to achieve our goals. We will only use the natural nutrients that are already there.

6. Don't eat much seafood? So you may think that none of this affects you. If you eat organic fruits and vegetables, or almost any kind of meat, it does. A major fertilizer used on organic fruits and vegetables is fish fertilizer. Fish fertilizer is made from fish byproducts. Most livestock feed is heavily

supplemented by fish protein, once again from fish byproducts.

7. The usual growing season for most phytoplankton is in the winter; because of the summer ocean boundary layer.

8. The thermal boundary layer is characterized by warm water above, and cold water below. This is usually referred to as the thermocline.

9. The summer thermocline blocks what would be a much better growing season for the phytoplankton. Nutrient depletion drops the summer production of phytoplankton to about a fiftieth of what it should be.

10. Phytoplankton absorb relatively large amounts of carbon dioxide. Carbon dioxide is a molecule made up of one atom of carbon and two atoms of oxygen. If we remove the one atom of carbon we are left with two atoms of oxygen (02). This is easily achieved using phytoplankton photosynthesis.

11. In essence, what we're doing is using solar energy (Photosynthesis) to convert the carbon dioxide into oxygen, and a stable form of carbon.

12. The world's oceans regularly absorb and expel atmospheric gases. You might say the oceans breathe. they breathe much more oxygen out then they breathed in.

13. When phytoplankton die, whether or not they are eaten, they eventually carry most of their carbon to the sea floor.

14. If we increase the amount of phytoplankton in the ocean, we also increase all of the other sea life.

15. The thermocline has good and bad aspects. It keeps us from being plunged into another Ice Age, but it also retards the summer growth of phytoplankton.

16. In a properly managed algae bloom we would get at least 1 lb. Of algae per square-foot of ocean.

17. Phytoplankton only live a day or two before dying and carrying their carbon to the bottom of the ocean.

18 Phytoplankton double in number every day if all of their growing conditions are met.

19. The summer, one shot, single spot, re-mineralization of the surface of the ocean would probably only last about 20 to 25 days.

20. 50% of the world's legal fish harvest comes from just one-point two million square miles of the ocean. This is only equal to less than one half the surface area of Canada.

21. The oceans make up over 129,000,000 square miles. Less than 1% of this provides over half the fish we consume.

22. The 1% of the ocean that provides half our seafood are known as areas of natural upwell. The grand Banks is one example of these areas.

23. Each year these areas of natural upwell also reduce carbon dioxide by between one hundred and forty billion tons, and one hundred and sixty billion tons.

24. Phytoplankton also reduce the Earth's temperature by producing sulfur particles that can help in the production of sun reflecting clouds.

25. The easiest way to provide nutrients to the phytoplankton is through the use of seabed re-engineering, or deep sea water wells.(More on seabed re-engineering later)

26. A deep seawater well is merely a tube that goes down below the thermocline and brings nutrient rich deep seawater to the surface for dispersal.

27. You can't just dump the deep seawater over the side, you must disperse it in a pattern as wide as possible because the cold deep seawater will resist mixing with the warmer surface water.

28. If we can match the areas of natural upwell, with artificial treatments, we can increase the amount of seafood available in the world by at least 50%. While we are increasing the

available seafood, we will also be dramatically decreasing our excess carbon dioxide.

29. Some of the phytoplankton are consumed by zooplankton. Both types of plankton are directly consumed by a multitude of larger, direct, and filter feeding creatures; including shellfish, shrimp, coral, and even some whales.

30. Zooplankton includes many different kinds of microscopic animals; including krill and even tiny jellyfish. Some of these zooplankton are eaten by slightly larger creatures; that are then eaten by larger creatures, until they get to the size that we consume as seafood.

Chapter 10; Mixing It All up

At this point you have all of the information necessary to figure out how YOU can defeat the thermocline, and feed the phytoplankton. (Sounds like a videogame) In the remainder of this book I will explain my plans for converting carbon dioxide into seafood.

Now, I would like you to sit back for a few minutes; and think. Ask yourself this question. How would I do it? What would be the most cost-effective way of doing it? Solar powered? Wind powered? Wave powered? Think outside the box, don't be afraid of dumb ideas, sometimes two dumb ideas can come together and make one smart idea. I'm sure some of your ideas will be better than some of mine.

Talk to your friends about this book. Explain the situation, and the problem. Invite them to come up with solutions. Only by working together can we come up with the best ideas. Think; how can defeat the summer thermocline, seriously reduce carbon dioxide, and drastically increase the amount of food available to the world's hungry people?

Mixing Method 1; Mixing Boats

One of the more labor-intensive and highly effective methods of fertilizing the surface of the ocean would be the use of mixing boats. Almost any old, ocean capable, 30 or 40 foot boat would do. In the beginning, small decommissioned fishing boats would work just fine. You would merely fit these boats with diesel driven pumps that would spray the water as far as possible to the sides and back of the boat. Slightly modified fireboats would work really well too. All you would have to do is adjust the fireboat's water intakes so they would suck water from a least 100 feet below. In the future I would expect that specially designed boats with lower operating costs would be built.

Each of these boats would be expected to cover an area of 50 miles by 1 mile each day, they would finish their 40X45 mile area in 32 working days, and then start over. To meet the requirements of the Virgin earth challenge only seven of these boats would be needed. They would need to operate for one year.

The total cost of operating the Seven mixing boats for one year would be approximately 4.2 million dollars.

Six of the boats would be in scheduled operation. The seventh would be maintained as a backup vessel for routine maintenance, and in case one of the others suffered a mechanical failure.

The total area covered by the six boats would be 10,800 Square miles. This would mean that each 32 day cycle would remove 60,000,000 tons of carbon. In ten, 32 day cycles, you would remove 600,000,000 tons of carbon from the ecosystem. That would exceed the Virgin earth challenge goal because the weight of the carbon in the carbon dioxide molecule is less than half of the total molecular weight. So in essence, removing 600,000,000 tons of carbon would mean you would actually be removing more then 1,200,000,000 tons of carbon dioxide from the atmosphere.

At the same time, you would be releasing over 600 million tons of oxygen into the biosphere. You must remember that what the plankton are actually doing, is converting the carbon dioxide molecules into stable carbon, and free oxygen. This would be an extremely profitable endeavor. It would produce billions of dollars worth of carbon credits.

Chapter 11; Going Medieval on carbon dioxide

We could if we wished, lower carbon dioxide levels back to what they were in the late 19th century. We could accomplish that by putting many more mixing boats into service.

Eight hundred, 30' or 40' mixing boats, could cover an area equal to the aforementioned 1,294,437 square miles of the world's oceans. Each boat would be responsible for mixing the water in the same 40 mile x 45 mile area. If each boat covered an area of 50 miles by 1 mile each day, they would finish their 40X45 mile area in 32 working days, and then start over. This would be done year-round in equatorial areas, but only for four to six months in more northern and southern latitudes. It wouldn't be necessary to have them all based at the same place, they could be scattered all over the world.

I'm nearly certain that when carbon credit accounting becomes mandatory, and worldwide, carbon credit units final value will be quite a bit more than my cost estimate of less than 1 cent per ton. At the time of this writing, carbon credits are trading for between $5 and $8 per unit. (One carbon credit unit equals 1 ton of carbon). Even if my calculations are off by a factor of 10, that would still only be a cost of 7 cents per ton.

Estimating an average $600,000 operating cost, per boat, per year. (A bit More in developed countries a bit less in underdeveloped countries) This would be a total operating expenditure, for all 800 boats, of $480 million per year. Each boat would be removing approximately 90,000,000 <u>tons</u> of carbon from the atmosphere each year. The entire fleet would be capable of removing 71,500,000,000 tons of carbon from the atmosphere per year. This would be a total of approximately 140,000,000,000 tons of carbon dioxide removed from the atmosphere every year. It must also be remembered that as well as consuming carbon dioxide, and in most cases falling to the bottom of the sea, the plankton would provide incredible amounts of food for the ocean's ecosystem. In five years, we would increase the available fish stocks on the planet by at least 100%. This would be on the order of 100,000,000 tons of additional seafood each year. This would amount to an additional harvest of 10 to 15 pounds for every single person on the planet.

Mixing Method 2; Wind Powered Mixing Buoys

You don't have to be a rocket scientist to figure out how these would work. They would be standard Marine buoys that would be fitted with the equipment to draw the deep water up to the surface and disperse it as far as possible, using one or more horizontal wind powered pumps. The deep-sea well (updraft tube), pumps, and spray nozzles would need to be made out of a material that would discourage clogging by Marine growth, possibly Teflon or polyethylene. These buoys would be anchored using very long, mooring cables. Ideally they would move in a circle of at least a mile as the wind pushed them in different directions. These buoys could be fabricated, and maintained locally in underdeveloped countries, with the idea in mind of enhancing the local fishery.

Mixing Method 3; Wind Powered Mixing Rafts or Barges

These rafts or barges would just be larger versions of the buoys. Each barge could have between 30 and 50 horizontal windmills on them spraying the water out as far as possible. Once again they could be fabricated locally with the key reason, of course, being local fisheries management. The buoys and barges wouldn't just fertilize the area directly under them. As the wind blows the surface of the ocean around, the buoys and barges would probably create a plume of plankton growth that could extend for 20 or 30 miles.

Mixing method 4; Offshore Drilling Platforms

This one is delightfully ironic. All offshore drilling rigs could be required to mix all the water within a given number of miles of the rig. Once they know the facts, the oil companies might very well do this voluntarily. They would realize that this requirement of reducing carbon and enhancing the local fisheries would make it much easier to get the okay to put these drilling platforms in. With the oil companies coming under more and more fire for the carbon dioxide problems, this would be worthwhile for them strictly on a public relations basis. Once again, this mixing would most likely create a wind driven plume that would extend far beyond the target area.

Mixing Method 5; Carrying Coal to Newcastle

Another way of re-mineralizeing the surface of the ocean
would be to carry the minerals out to sea in modified ore carrier
ships. The nutrients we are discussing here are for the most
part, minerals. A land-based source of these minerals would be
the dried silt that's dredged from almost all the world's rivers on
a regular basis. An ore carrier with the equipment to mix and
spray muddy silt laden ocean water out as far as a quarter of a
mile on each side. This method could be used, as a backup
method, wherever the deep ocean water was found to be less
nutrient rich. It could also be used to further enhance or expand
areas of natural upwell. This activity might be especially
worthwhile in areas of the Arctic, where the extra plankton
would produce extra clouds, and probably alleviate some of the
stress on the polar bear population.

Mixing Method 6; Wave Powered Mixing Rafts

This one should actually be referred to as swell powered mixing rafts. To understand how these would work, you need to, mentally, go down deep in the ocean and hover there in your imaginary wetsuit. If you hung a 100 ft. long rope off a raft, you would notice that the rope would move up and down between 10 and 20 ft., in relation to your position. The amount the rope would travel, would depend on how high the swells, or waves, were on the surface. If you were to grab hold of this rope, you would be pulled up strongly each time a swell, or wave moved under the raft. There would be many ways to use this free power to defeat the thermocline.

The simplest way the use this power would be to hang a heavy weight at the bottom of the rope. Next you would place a 6 to 10 ft. diameter disk every 8 to 10 ft. along the rope. As the rope moved up, and then was pulled back down by the weight, each of these disks would pull water up from the depths moving it toward the surface. The shape of the disks would need to be either a concave or conical, so that the water would be moved up. This method would probably be the easiest to build and maintain in Third World countries. The disks and rafts could actually be any size. You could also hang multiple ropes from each raft, depending on the raft's size.

These rafts would be anchored to a quarter mile long rope, or cable, so that as the wind blew in different directions, they would cover an area of up to half a mile

There would probably be five or six other ways that this up-and-down rope could be used, see what you can come up with.

Mixing method 7; Re-engineering the sea bottom

I've saved this method of defeating the thermocline for last for a reason. This method would require the most study, and commitment. We would literally be creating new areas of natural upwell. Most of the areas of upwell naturally defeat the thermocline, day after day, year after year, with no intervention by humans. You are already aware that this type of upwell is caused by a natural effect.

With the full commitment of all the parties involved, we could chose areas beneath some of the global deep-water currents and use shiploads of rock dropped to the bottom. This action would force a small portion of these currents toward the surface. Unlike the other methods suggested, once this was done there would be no need for any further maintenance or energy expenditure. Unfortunately, doing this would be rather permanent, so it would need to be done very carefully.

A less permanent and more easily reversed solution, might be the use of concrete pipe elbows. These huge concrete pipes would catch a small portion of the current and redirect it toward the surface. These pipes could be as much as 30 to 50 ft. in diameter. Each of these pipes would probably continuously treat a 5 to 10 square-mile area of the Upper Ocean. This method would also require little or no maintenance.

REVIEW II

At this point you should understand The Following.

1. Mixing boats would be 30 to 60 ft. ocean going boats, or ships. These vessels would be equipped with long flexible hoses, and pumps, that would pull nutrient rich water from below the thermocline, and spray it on the surface.

2. Mixing buoys would be standard marine buoys that would be fitted with the equipment to draw the deep water up to the surface and disperse it as far as possible, using one or more horizontal wind powered pumps.

3. We could if we wished, lower carbon dioxide levels back to what they were in the late 19th century. We could accomplish that by putting many more mixing boats or devices into service.

4. Wind powered mixing rafts or barges, would just be larger versions of the buoys. Each barge could have between 30 and 50 horizontal windmills on them spraying the water out as far as possible.

5. All offshore drilling rigs could be required to mix all the water within a given number of miles of the rig. Once they

know the facts, the oil companies might very well do this for positive public relations.

6. We could re-mineralize the ocean sunlight zone by carrying the minerals out to sea in modified ore carrier ships. The nutrients we are discussing here are, for the most part, minerals. A land-based source of these minerals would be the dried silt regularly dredged from Inland rivers.

7. Wave, or swell powered rafts would probably offer the second least labor-intensive method of breaking the thermocline. This method would need very little maintenance. These would probably be the easiest to build, and maintain, in underdeveloped countries.

8. The method requiring the least amount of maintenance would be seabed re-engineering. This method could last hundreds or even thousands of years and require absolutely no maintenance. The only problem is once instituted it would be reasonably permanent. It would need to be done very carefully.

9. A less permanent and more easily reversed solution, might be the use of concrete pipe elbows. These huge concrete pipes would catch a tiny portion of a deep ocean current, and redirect it toward the surface. These pipes could be as much as 30 to 50 ft. in diameter. They could be easily moved or destroyed if it

was found that they were no longer necessary.

10. there is really no best way to break the thermocline. Each ocean locale would probably require a different approach. Most of these mixing devices could be built using recycled materials.

Part 3;

Chapter 12; Let's Not Forget about the Lakes

Lakes are very significant in the removal of carbon dioxide from our atmosphere. Freshwater, and salt water lakes, cover an area of more than one and half million square miles on planet Earth.

If you've ever gone swimming in a lake and found that the bottom was covered with a black goop, you have seen the carbon that the lake is removing. That black substance is probably made up of 70 to 80% carbon.

The algae in the Lake has gone through its life-cycle and taken the carbon into itself. When the algae died, it fell to the floor of the Lake carrying the carbon with it. That carbon will probably lay on the lake bottom for hundreds or even thousands of years.

In the early spring you can only see down 4 or 5 ft. in most lakes. By midsummer you can usually see 15 to 20 ft. down. The reason is our good old friend the thermocline.

The algae in most lakes is pretty close to phytoplankton in its activity and life-cycle. The surface area of the lakes here on

Earth is pretty close to the areas of natural up-well in the oceans. Lakes, for the most part, can be thought of as forests. Lakes sequester about three times, the carbon dioxide of a healthy forest of equal size.

The Great Lakes alone have a surface area of ninety four thousand square miles.

Commercial fishermen on the Great Lakes are having a bit of a problem. The Whitefish industry on the lakes has discovered that, though the fish they're taking are of proper length, the weight of each fish is down significantly. Part of that is due to global warming. Lake Superior's surface temperature has risen 4.5°F since 1979. This means that the thermocline on the lakes is forming several weeks earlier, and breaking down several weeks later then it was 30 years ago. It seems apparent that the fish are not finding enough to eat.

The commercial and sports fishing industry on the Great Lakes is worth about $4 billion per year. Until I started doing research for this book I had no idea that there was a commercial fishing industry on the Great Lakes.

There doesn't seem to be a lot of disagreement on the fact that global warming is happening. The disagreement seems to stem from whether humans are causing it, or if it's part of a natural cycle. I happen to believe that it's part of the natural cycle. I also believe that by feeding the phytoplankton and algae, we can mitigate the potential damages, and produce millions of tons of extra fish and seafood.

Chapter 13; Feeding Lake Algae

The same methods used for breaking the thermocline in the ocean, can be used to break up the thermocline in large lakes. The good thing about working with lakes is that you're much closer to shore. On the ocean there's no need to feed the phytoplankton within about 5 miles of the coast. The wave action against the beach, and tidal currents, churn up the mineral nutrients that the phytoplankton need.

Most lakes, on the other hand, have little, or no, wave action. Inland bodies of water that are less than 30 ft. deep seldom require nutrient mixing. When the thermocline reaches the bottom of the shallow lakes, normal convection mixes the nutrients naturally. All lakes that are more than 30 ft. deep can be helped, using the same methods used in the ocean.

The way to figure out how your favorite lake is doing is quite simple. Just look down. If you can see 20 ft. down into the water, it means the lack of nutrients has forced the algae into its summer dormant stage. Humans seem to like lakes that are clear, but this clarity is the leading indication that the lake is not really doing that well.

The good thing about lakes, is that the nutrient mixing can

usually be carried out from a dock, or from shore. If you would like to try a little experiment, try this. All you'll need is a gasoline, or electric driven pump; a couple of hundred feet of hose; and a sprinkler.

Run the intake hose from the pump out into deep water. Hook the sprinkler up to the output hose on the pump. Put the sprinkler at the end of a dock, or float of some kind. Start the pump. Run the pump for an hour or two every day for about a week. You'll see a marked difference in the amount of algae in the area of the sprinkler. After a few weeks you will probably find this area offers much better fishing then other areas of the lake.

Each square-mile of Lake probably sequesters about five thousand tons of carbon each year naturally. If these bodies of freshwater were mixed properly, they could probably sequester as much as 15,000 tons of carbon.

Chapter 14; Show Me the Money

Earlier in this book I promised you that people participating in global warming abatement would reap financial, as well as moral rewards. Here's how it will probably work.

In Europe, in 2005, they instituted something called carbon cap and trade. In Europe this cap and trade scheme covers over 10,000 carbon dioxide producing installations. In essence what happens is the government sets carbon dioxide emission limits. If a facility exceeds its carbon dioxide limits the owners must purchase carbon credits from installations that have managed to lower their carbon dioxide emissions. If they don't managed to do this they pay a fine to the government. An open market has evolved somewhat like the commodities markets here in the US. The overall scheme is very complicated, much to complicated to go into in this small book. If you want more information you can do a Google search using the keywords; **"European cap and trade."**

personally I feel that this cap and trade foolishness is just another governmental grab for power and money.

So what does this mean to you?

These cap and trade scenarios allow for something called

Carbon offsets. This is where the phytoplankton comes in. If you can prove to an independent entity that you are sequestering carbon dioxide, they will provide you with carbon credits. Each carbon credit is equal to one ton of sequestered carbon dioxide. These carbon credits can be sold on the open market to companies that need them, or people who just want to offset their drive to work.

The current market value of these carbon credits (summer of 2011) is running between $20 and $24 each. If you can prove you sequestered 1000 tons of carbon dioxide you just made $20,000 to $24000. if you can prove you sequestered one million tons of carbon dioxide, you just made 22 to 24 million dollars.

There will also be a lot of money made in investing in carbon credits. If today's market value of one carbon credit is $20 and you buy it, and the price goes up, you sell it and make a profit. These carbon credits are currently being traded on various commodities exchanges, just like wheat; or gold; and silver.. I don't recommend this activity for people who are unschooled in the business, because the price could possibly go down.

The US government hasn't instituted cap and trade yet, but when they realize how much money, and control, the other world governments are gaining, I'm sure we will see it here by 2013. The Environmental Protection Agency is already making all kinds of anti-carbon noises. Several US states have begun a voluntary cap and trade. These voluntary systems will most likely not work. But these so-called voluntary systems will lay the groundwork for a mandatory system.

I personally feel that human caused global warming, is just a bunch of junk science. I also don't believe that all of this carbon foolishness will have any effect at all. But if enough people believe it, we will all suffer the consequences. Once again I have to reiterate that the main reason I'm behind this phytoplankton cultivation, is to revive and replenish our lakes and oceans.

As you read the previous sentence, one more person died of starvation somewhere in the world. The really sad thing is, there's an 80% chance that it was a small child who died.

I know that it may seem callous to use carbon dioxide and global warming as a method when it comes to starving people. But if the global warming people unite with the anti-starvation people, we will become a power to contend with. It's truly a case of the end justifying the means. So even if you don't believe in global warming, I'm sure you do believe that there are people out there starving to death every minute and a half.

Remember that 60 to 80% of those starving people are small children. Let them eat fish.

Chapter 15; Influence the weather?

I stumbled across some interesting information as I was doing research for this book. A hurricane can only form, and grow, if the surface water temperature near it is above 79° F. A hurricane will begin losing power at a surface water temperature of 78°F, and will literally begin dying when the surface water temperature hits 75° F.

I had absolutely no idea that hurricanes were this delicate. A fleet of 100 mixing boats could be deployed in front of an approaching hurricane and knock its power down significantly.

The water that the mixing boats draw from 100 ft. below is much colder than the surface water. When doing the calculations for the removal of carbon dioxide I never took into consideration, the fact that we would actually be cooling the surface of the ocean. This is a very interesting development.

There is a possibility that lowering water temperature in a large area might create a localized low-pressure area and cause hurricanes to move in the direction of the cooler water. This little quirk of weather could make it possible for us to actually steer hurricanes. If we could influence hurricanes to move in a northerly direction, they would hit water that was naturally cooler and dissipate their strength all by themselves.

More than you probably want to know about hurricanes

Atlantic hurricanes usually begin as complicated groups of thunderstorms off the west coast of Africa. For hurricanes to grow, ocean waters MUST be above 26°C (79°F). Below this magic temperature, hurricanes will not form or grow.

Hurricanes will lose ferocity rapidly once they move over water below this threshold temperature. Ocean temperatures in the tropical East Pacific and the tropical Atlantic routinely surpass this threshold in the spring and summer. (Hurricane season) In the summer the waters between Africa and the United States routinely rise to as high as 85° Fahrenheit, and occasionally as high as 88° Fahrenheit.

Pacific hurricanes come in from the opposite direction. If you look at a global summer water temperature map, you will see that the water temperature in both areas are conducive to hurricane activity. There are many other factors involved in the formation of hurricanes, but the factor we're interested in is surface water temperature.

If you lower the temperature of the surface of the ocean to 78° or 77° you will be, in effect, damaging the hurricane. If we lower the water temperature just 7° or 8° in front of a hurricane, we will lower the ferocity of the hurricane by several category numbers.

There's also the possibility, that if we lower the surface temperature of the water, in front of a hurricane, in a large enough area, we could actually kill a hurricane. Much more research needs to be done in this area. Could this be the beginnings of whether control?

FINAL REVIEW

At this point you should understand The Following.

1. There are as many species of phytoplankton in the ocean as there are different kinds of plants on land.

2.There are as many species of zooplankton in the ocean as there are different kinds of insects on land.

3. We are only engaging in overfishing because it is necessary to sustain the lives of millions of human beings. This book how global fishing can be tripled or quadrupled, and at the same time leave the oceans, and our atmosphere healthy.

4. phytoplankton are necessary for the existence of all of the ocean's creatures.

5. We won't introduce any chemicals or fertilizer into the oceans to achieve our goals. We will only use the natural nutrients that are already there.

6. Don't eat much seafood? So you may think that none of this affects you. If you eat organic fruits and vegetables, or almost any kind of meat, it does. A major fertilizer used on organic fruits and vegetables is fish fertilizer. Fish fertilizer is made from fish byproducts. Most livestock feed is heavily supplemented by fish protein, once again from fish byproducts.

7. The usual growing season for most phytoplankton is in the winter; because of the Summer Ocean thermal boundary layer.

8. The thermal boundary layer is characterized by warm water above, and cold water below. This is usually referred to as the thermocline.

9. The summer thermocline blocks what would be a much better growing season for the phytoplankton. Nutrient depletion drops the summer production of phytoplankton to about a fiftieth of what it should be.

10. Phytoplankton absorb relatively large amounts of carbon dioxide. Carbon dioxide is a molecule made up of one atom of carbon and two atoms of oxygen. If we remove the one atom of carbon we are left with two atoms of oxygen (02). This is easily achieved using phytoplankton photosynthesis.

11. In essence, what we're doing is using solar energy (Photosynthesis) to convert the carbon dioxide into oxygen, and a stable form of carbon.

12. The world's oceans regularly absorb and expel atmospheric gases. You might say the oceans breathe. they breathe much more oxygen out then they breathed in.

13. When phytoplankton die, whether or not they are eaten, they eventually carry most of their carbon to the sea floor

14. if we increase the amount of phytoplankton in the ocean, we also increase all of the other sea life.

15. The thermocline has good and bad aspects. It keeps us from being plunged into another Ice Age, but it also retards the summer growth of phytoplankton.

16. In a properly managed algae bloom we would get at least 1 lb. Of algae per square-foot of ocean.

17. Phytoplankton only live a day or two before dying and carrying their carbon to the bottom of the ocean.

18 Phytoplankton double in number every day if all of their growing conditions are met.

19. The summer, one shot, single spot, re-mineralization of the surface of the ocean would probably only last about 20 to 25 days.

20. 50% of the world's legal fish harvest comes from just one-point two million square miles of the ocean. This is only equal to less than one half the surface area of Canada.

21. The oceans make up over 129,000,000 square miles. Less than 1% of this provides over half the fish we consume.

22. The 1% of the ocean that provides half our seafood are known as areas of natural upwell. The grand Banks is one example of these areas.

23. Each year these areas of natural upwell also reduce carbon dioxide by between one hundred and forty billion tons, and one hundred and sixty billion tons.

24. Phytoplankton also reduce the Earth's temperature by producing sulfur particles that can help in the production of sun reflecting clouds.

25. The easiest way to provide nutrients to the phytoplankton is through the use of seabed re-engineering, or deep sea water wells.

26. A deep seawater well is merely a tube that goes down below the thermocline and brings nutrient rich deep seawater to the surface for dispersal.

27. You can't just dump the deep seawater over the side, you must disperse it in a pattern as wide as possible because the cold deep seawater will resist mixing with the warmer surface water.

28. If we can match the areas of natural upwell, with artificial treatments, we can increase the amount of seafood available in the world by at least 50%. While we are increasing the available seafood, we will also be dramatically decreasing our excess carbon dioxide.

29. Some of the phytoplankton are consumed by zooplankton.

Both types of plankton are directly consumed by a multitude of larger, direct, and filter feeding creatures; including shellfish, shrimp, coral, and even some whales.

30. Zooplankton includes many different kinds of microscopic animals; including krill and even tiny jellyfish. Some of these zooplankton are eaten by slightly larger creatures; that are then eaten by larger creatures, until they get to the size that we consume as seafood.

31. Mixing boats would be 30 to 60 ft. ocean going boats, or ships. These vessels would be equipped with long flexible hoses, and pumps, that would pull nutrient rich water from below the thermocline, and spray it on the surface.

32. Mixing buoys would be standard marine buoys that would be fitted with the equipment to draw the deep water up to the surface and disperse it as far as possible, using one or more horizontal wind powered pumps.

33. We could if we wished, lower carbon dioxide levels back to what they were in the late 19th century. We could accomplish that by putting many more mixing boats or devices into service.

34. Wind powered mixing rafts or barges, would just be larger versions of the buoys. Each barge could have between 30 and 50 horizontal windmills on them spraying the water out as far

as possible.

35. All offshore drilling rigs could be required to mix all the water within a given number of miles of the rig. Once they know the facts, the oil companies might very well do this for positive public relations.

36. We could re-mineralize the ocean sunlight zone by carrying the minerals out to sea in modified ore carrier ships. The nutrients we are discussing here are, for the most part, minerals. A land-based source of these minerals would be the dried silt regularly dredged from Inland rivers.

37. Wave, or swell powered rafts would probably offer the second least labor-intensive method of breaking the thermocline. This method would need very little maintenance. These would probably be the easiest to build, and maintain, in underdeveloped countries.

38. The method requiring the least amount of maintenance would be seabed re-engineering. This method could last hundreds or even thousands of years and require absolutely no maintenance. The only problem is once instituted it would be reasonably permanent. It would need to be done very carefully.

39. A less permanent and more easily reversed solution, might be the use of concrete pipe elbows. These huge concrete pipes

would catch a tiny portion of a deep ocean current, and redirect it toward the surface. These pipes could be as much as 30 to 50 ft. in diameter. They could be easily moved or destroyed if it was found that they were no longer necessary.

40. there is really no best way to break the thermocline. Each ocean locale would probably require a different approach. Most of these mixing devices could be built using recycled materials.

41. Freshwater, and salt water lakes cover more than one half-million square miles on planet Earth. Not all of our lakes develop a thermocline, but the ones that do can be handled the same way we handle the oceans.

42. The black goop you sometimes find, on the bottom, in the shallow waters of a Lake, is the Lake algae after it has completed its lifecycle. That black goop is mostly scavenged carbon. That carbon has been removed from atmospheric carbon dioxide molecules. This action has, of course, also released a large amount of oxygen.

43. The distance you can see down into a Lake is an indicator of the lake's health status. If you can see down more than 20 ft. that Lake has developed a thermocline. This thermocline will have an adverse affect on fishing in that particular Lake.

44. A healthy Lake will sequester about three times the carbon

dioxide that a healthy Forest of the same size will.

45. The Great Lakes, in the United States, have a surface area of ninety four thousand square miles. The fishing industry on the Great Lakes is worth about $4 billion a year.

46. The surface temperature of Lake Superior has risen 4.5° since 1979. This means the thermocline is forming almost a month earlier, and lasting almost a month longer, than it was 30 years ago.

47. The fish in Lake Superior bar of about the same length that they were 30 years ago; but they're about 30% lighter. This is having a negative effect on the commercial fishing in the Great Lakes.

48. The same methods can be used to break the thermocline in lakes as we use on the ocean. The good thing about working with lakes is we're almost always much closer to shore, and the electrical power grid.

49. Inland bodies of water that are less than 30 ft. deep seldom require nutrient mixing. When the thermocline reaches the bottom of shallow lakes, normal warm water convection mixes the nutrients; even in summer.

50. All lakes that are more than 30 ft. deep can be helped using the same methods that we use on the ocean. We all like a nice

clear Lake, but that clarity is an indication of the lakes poor health.

What can you do?

Talk about this book, or at least the plan in this book. Get a buzz going on the Internet. Add information in this book to your face book pages. Ideas with little or no money behind them, can become huge through the power of word-of-mouth. Each person you tell about this book will probably tell others. The ideas will spread, and become popular. This will also be a way to stop, or at least curtail, the coming cap and trade. Once the politicians start smelling the extra tax money, and power, that cap and trade will produce, there will be no turning back. That extra money will come out of your pocket as well as mine. If you care about global warming, world hunger, or your pocketbook, get busy, the ball is in your court.

==

Online Bibliography, And Data Verification Sites

================================

This link to a page on the website of the United States environmental protection agency verifies that phytoplankton is in fact, 40% carbon, also explains that phytoplankton removes other undesirable compounds from the atmosphere and ocean as well. *Scroll down to 2.2*

http://www.epa.gov/greatlakes/invasive/zmussels/sec2.html

This link to a page that verifies the passage of gases from the atmosphere to the ocean, and from the ocean to the atmosphere.

Read full page

ahttp://www.atmosphere.mpg.de/enid/1w0.html

This link to a page on the NASA website verifies that phytoplankton are apparently capable of creating their own cloud cover. *Scroll down six to 8 inches* http://www.nasa.gov/lb/centers/goddard/news/topstory/2004/0702planktoncloud.html

This link to another page on the NASA website (Earth Observatory) verifies the fact that plankton doubles numerical population each day.. ***Read full page***

http://earthobservatory.nasa.gov/Library/Phytoplankton/phytoplankton2.html

This link to another page on the NASA website (Earth Observatory) verifies the following facts; The natural upwell all of the nutrients in winter, plankton live for a day or two, and the ocean food chain. ***Read full page, but pay special attention to the bottom third.***

http://earthobservatory.nasa.gov/Library/Phytoplankton

The NASA link below verifies the following fact; under ideal conditions phytoplankton double in number each day. It is also a good page to read for your own information. It verifies almost everything that I've said in this book.

http://science.nasa.gov/earthscience/oceanography/living

-ocean/ocean-color

This link is a page on the coastal Conservancy website, that gives a good explanation of upwelling, and areas of the ocean that are the most productive. This page also explains the fact that 50% of the fish harvest of the world comes from only 1% of the oceans. ***Read bottom half of page, and left sidebar***

http://www.coastalconservancy.ca.gov/coast&ocean/summer98/a02.htm

This link is a page explaining carbon credits, and how they work. ***Read as much as you're interested in, it's Strictly FYI***

http://www.worldchanging.com/archives/002864.html----

Page that explains what's necessary for a hurricane to form or continue to grow.

http://www.theweatherprediction.com/tropical/

If you go to the page below it will show a temperature gradient map that shows why hurricanes on the the American southern Atlantic Coast travel to the United States from the west coast of Africa.

You will also notice that about 100 mi. below the American Mexico border the same temperature gradient appears in the Eastern Pacific. All of the orange areas will allow hurricanes to grow in ferocity.

http://www.wunderground.com/tropical/

Use the link below if you would like to read a free sample of the first 11 chapters of my latest action, comedy, future fiction, novel.

http://www.theshancreekgang.com/